# SpringerBriefs in Electrical and Computer Engineering

More information about this series at http://www.springer.com/series/10059

Zhi Wang • Jiangchuan Liu • Wenwu Zhu

# Social Video Content Delivery

Zhi Wang
The Graduate School at Shenzhen
Tsinghua University
Shenzhen, Guangdong, China

Jiangchuan Liu
School of Computing Science
Simon Fraser University
Burnaby, BC, Canada

Wenwu Zhu
Department of Computer Science
     and Technology
Tsinghua University
Beijing, China

ISSN 2191-8112                      ISSN 2191-8120   (electronic)
SpringerBriefs in Electrical and Computer Engineering
ISBN 978-3-319-33650-3          ISBN 978-3-319-33652-7   (eBook)
DOI 10.1007/978-3-319-33652-7

Library of Congress Control Number: 2016939413

Printed on acid-free paper

This Springer imprint is published by Springer Nature
The registered company is Springer International Publishing AG Switzerland

# Preface

Recently, the world has witnessed the convergence of online social network services and online video services; users import videos from content-sharing sites and propagate them among their social connections by re-posting them. Online social networks have reshaped how multimedia content is generated, distributed, and consumed on the Internet today. Given the massive amount of user-generated content shared in online social networks, users are now engaged as active participants in the social ecosystem rather than as passive receivers of media content. This revolution is being driven further by the deep penetration of 3G/4G wireless networks and smart mobile devices that are seamlessly integrated with online social networking and media-sharing services.

Despite increasingly abundant bandwidth and computational resources, the ever-increasing volume of data created by user-generated video content—along with the boundless coverage of socialized sharing—presents unprecedented challenges. In this book, we first present the challenges in social-aware video delivery. Then, we present a primary framework for social-aware video delivery and a thorough overview of the possible approaches. Moreover, we identify the unique characteristics of social-aware video access and social content propagation and reveal, in detail, the design and integration of individual modules that are aimed at enhancing user experience in the social network context. Finally, we present future research directions and discussions.

Shenzhen, Guangdong, China      Zhi Wang
Burnaby, BC, Canada      Jiangchuan Liu
Beijing, China      Wenwu Zhu

# Acknowledgments

We would like to express our gratitude to our colleagues, friends, and our families, who saw us through this book; to the editors and reviewers; and to all those who provided support and discussed things over. Particularly, we would like to thank Professor Sherman Shen and Springer for enabling us to publish this book. We also thank the research grant agencies/programs that have provided enormous support to our work; in particular, NSFC for Z. Wang, Canada NSERC for J. Liu, and National Basic Research Program of China for W. Zhu, respectively.

# Contents

# Chapter 1
# Introduction

## 1.1 Background

Over the past decade, we have witnessed a rapid evolution toward a new generation of networked, shared multimedia in the Web 2.0 era. Today, high-definition, 3D and multi-view videos can be readily captured and viewed by personal computing devices, and conveniently processed and stored with remote cloud resources. Despite the increasingly abundant bandwidth and computational resources, the ever-increasing data volume of user-generated video content and the boundless coverage of socialized sharing present unprecedented challenges to both content and network service providers. The highly diverse content origins and distribution channels further complicate the design and management of online video sharing systems. Here, we present a state-of-the-art survey of social-aware video delivery, identify key issues in this promising field, and present solutions.

Online social network services connect users through "friending" (e.g., Facebook), "following" (e.g., Twitter), or professional connections (e.g., LinkedIn). Such applications have successfully changed how people connect to each other and how they share information, including videos. Recently, the convergence of online social network services and online video services allows users to import videos from video sharing sites to their online social networks, where those videos are propagated along the social connections between people who re-share them. Social behaviors have dramatically reshaped how videos are disseminated to users: Specifically, many people now receive videos directly from their friends. For example, the online music video "Gangnam Style" attracted over one billion views within 6 months after it was uploaded, due to its propagation over popular online social networks including Twitter and Facebook. In 2015, the duration of videos watched and shared every day by Facebook users totals approximately 500 years, and over 700 videos are shared on Twitter every minute [26].

© The Author(s) 2016

Z. Wang et al., *Social Video Content Delivery*, SpringerBriefs in Electrical and Computer Engineering, DOI 10.1007/978-3-319-33652-7_1

Conventional video delivery strategies, e.g., the original client/server streaming, IP and application-layer multicast/peer-to-peer, and content delivery networks (CDNs), mainly focus on improving the network delivery performance to meet the increasing scale of video requests. These strategies have generally assumed that the content comes from centralized service providers and that users only *passively* receive the content [23]. For videos that are shared over social networks, however, the access patterns are much more dynamic and are affected by individuals and their social behaviors during propagation [18]. A more sophisticated system design for social media content delivery is thus in demand.

## 1.2   Previous Efforts

Online social networks have become popular Internet services. Based on logs from Flickr, LiveJournal, and Orkut, Mislove et al. studied the topology of the social graph and confirmed the power-law, small-world, and scale-free properties [15]. Krishnamurthy et al. investigated Twitter and identified distinct classes of Twitter users and their behaviors [2].

In online social networks, content spreads among users. A great deal of research effort has been devoted to studying how information propagates through online social networks. Kwak et al. [11] investigated the impact of users' retweets on information diffusion in Twitter. Dodds et al. used an epidemic model to study information propagation by considering each piece of information as an infective disease that spreads via social connections [7]. Kempe et al. investigated how to maximize the spread of influence in an online social network [10], and Hartline et al. utilized this understanding to achieve revenue maximization, by selecting the best set of initial users to push the information. Domingos et al. assessed the value of social networks in estimating the potential buyers of a product or a service, which can be influenced by existing customers [8].

Many architectures have been proposed for large-scale content service systems, including (1) server-based, e.g., CDN and cloud-based methods [17], (2) client-based, e.g., P2P content delivery [13], and (3) hybrid, e.g., a CDN coupled with a P2P delivery framework [25]. For Internet-scale social content services, replicating the content across different geographic regions is a promising approach for providing good-quality services to users [1]. Zhu et al. proposed allocating cloud servers at the network edges to distribute multimedia content to users [27].

*Content-Based Replication* From the content aspect, traditional content delivery primarily concerns itself with content popularity, and it allocates storage and bandwidth resources accordingly [9]. Since 2005, a huge amount of content has been generated by users. Exploring the "social networks" of content can effectively help users fetch content precisely [5], i.e., content to be requested by users in the future. However, online social networks have greatly changed the assumptions of traditional replication algorithms [3], e.g., the distribution of content is shifted from a "central-edge" manner to an "edge-edge" manner, resulting in the hybrid long-tailed and close-to-uniform popularity distribution. Li et al. studied content sharing and observed a skewed content popularity distribution and the "power-law" activity intensity distribution of users [12]. To better serve such social content, some social-aware content replication schemes have been proposed.

*Social-Based Replication*  User relationships and influences were studied in online social networks. After a person shares a given item, that person's friends may also be attracted to request it. Pujol et al. investigated the difficulties of scaling an online social network hosting system and designed social partitioning and replication middleware, in which users' friends' data can be co-located in the same server [18]. Tran et al. also studied the partitioning of user-generated content by taking social relationships into consideration [20]. Nguyen et al. investigated how to improve the efficiency of a social media system in cases of server failure, by using the social locality pattern [16]. Wang et al. observed that a social network can be used to predict content access patterns in a standalone content sharing system, e.g., Youku [21]. Wu et al. studied how to minimize the cost of migrating social media among servers in different regions [24]. Cheng et al. studied using social media partitions to balance the server load and preserve the social relationship [6].

## 1.3  Challenges in Social Video Delivery

Today's social video content delivery systems are facing the following challenges.

*Users, Not Service Providers, Determine How Videos Reach New Audiences*  First, in an online social network, content is generated, propagated, and disseminated by users. In 2014, YouTube reported that users uploaded over 100 h of video clips every minute. Content delivery systems thus have to distribute a much larger volume of user-generated videos than what has ever been handled by conventional content providers [4]. Second, users share videos through social connections, and they tend to receive videos from their friends. Consequently, service providers no longer have tight and centralized control over the dissemination of content.

*Dynamic Content Propagation*  Social propagation is affected by a combination of factors, including social topology, user behaviors, and the inherent content characteristics [22]. Because it has many influencing factors, propagation is highly dynamic, and traditional content delivery strategies generally lack predictive tools to infer such inherent dynamics.

*Changes in the Popularity of Social Content*  Changes in content origins and patterns of social propagation also change the popularity distribution of videos. On one hand, the overall skewness of social video popularity distribution has been amplified; on the other hand, certain initially unpopular videos can be rediscovered and then shared among users with close social relationships. Video popularity is a key factor in designing and optimizing video delivery systems, and the change thus has strong implications. For example, the request hit ratio can be degraded by over 70 % when traditional cache strategies are used to handle online social content [14].

The increasing popularity of online social network services has fundamentally changed the landscape of online video content delivery and requires a new set of content delivery strategies, including content delivery overlay construction, network resource allocation, and content deployment. In this book, we present these and other strategies that can significantly improve video delivery quality within the context of the convergence of online social network services and online video services.

## 1.4  Social-Aware Video Delivery Framework

The challenges described above demand the joint study of user behavior, social video popularity, and social propagation to improve the performance of content delivery in the social network context. To this end, Fig. 1.1 illustrates a general framework of social-aware video content delivery strategies. In this framework, we collect social (e.g., social relationship and behaviors), content (e.g., types and features), and context (e.g., time and location) information to study popularity and propagation characteristics. Machine learning models have been designed to predict the evolution of social video popularity and the propagation patterns (e.g., the size of a social cascade). The prediction models estimate how videos will be shared in different locations, and network resources can then be dynamically allocated among these locations. In addition, content replication and caching strategies can be incorporated to further improve the efficiency with which the content is shared [19].

This work explores the unique characteristics of social-aware video access and the social propagation of content and closely examines the design and integration of individual modules into the framework.

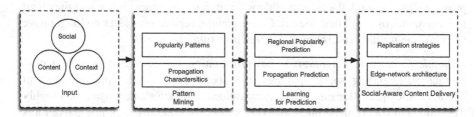

**Fig. 1.1**  Framework of social-aware video content delivery

# References

1. V.K. Adhikari, S. Jain, and Z.L. Zhang. "Where Do You Tube? Uncovering YouTube Server Selection Strategy". In: *IEEE International Conference on Computer Communications and Networks*. 2011.
2. Martin Arlitt Balachander Krishnamurthy Phillipa Gill. "A Few Chirps About Twitter". In: *ACM Workshop on Online Social Networks (WOSN)*. 2008.
3. F. Benevenuto et al. "Characterizing User Behavior in Online Social Net- works". In: *ACM Internet Measurement Conference (IMC)*. 2009.
4. M. Cha et al. "I Tube, You Tube, Everybody Tubes: Analyzing the World's Largest User Generated Content Video System". In: *ACM SIGCOMM*. 2007.
5. X. Cheng and J. Liu. "Load-Balanced Migration of Social Media to Content Clouds". In: *ACM Network and Operating System Support for Digital Audio and Video (NOSSDAV)*. 2011.
6. X. Cheng and J. Liu. "NetTube: Exploring Social Networks for Peer-To-Peer Short Video Sharing". In: *IEEE International Conference on Computer Communications (INFOCOM)*. 2009.
7. P.S. Dodds and D.J. Watts. "A Generalized Model of Social and Biological Contagion". In: *Journal of Theoretical Biology* 232.4 (2005), pp. 587–604.
8. P. Domingos and M. Richardson. "Mining the Network Value of Customers". In: *ACM SIGKDD Conference on Knowledge Discovery and Data Mining (KDD)*. 2001.
9. J. Kangasharju, J. Roberts, and K.W. Ross. "Object Replication Strategies in Content Distribution Networks". In: *Computer Communications* 25.4 (2002), pp. 376–383.
10. D. Kempe, J. Kleinberg, and . Tardos. "Maximizing the Spread of Influence Through a Social Network". In: *ACM SIGKDD Conference on Knowledge Discovery and Data Mining (KDD)*. 2003.
11. H. Kwak et al. "What Is Twitter, a Social Network or a News Media?" In: *ACM International Conference on World Wide Web (WWW)*. 2010, pp. 591–600.
12. Haitao Li, Haiyang Wang, and Jiangchuan Liu. "Video Sharing in Online Social Network: Measurement and Analysis". In: *ACM Network and Operating System Support on Digital Audio and Video Workshop (NOSSDAV)*. 2012, pp. 83–88.
13. Y. Liu, Y. Guo, and C. Liang. "A Survey on Peer-To-Peer Video Streaming Systems". In: *Peer-to-peer Networking and Applications* 1.1 (2008), pp. 18–28.
14. A. Mislove. "Rethinking Web Content Distribution in the Social Media Era". In: *NSF Workshop on Social Networks and Mobility in the Cloud*. 2012.
15. A. Mislove et al. "Measurement and Analysis of Online Social Networks". In: *ACM Internet Measurement Conference (IMC)*. 2007.
16. K. Nguyen et al. "Preserving Social Locality in Data Replication for Social Networks". In: *IEEE International Conference on Distributed Computing Systems (ICDCS) Workshop on Simplifying Complex Networks for Practi- tioners*. 2011.
17. G. Peng. "CDN: Content Distribution Network". In: *arXiv preprint cs/0411069* (2004).
18. J.M. Pujol et al. "The Little Engine(s) That Could: Scaling Online Social Networks". In: *ACM SIGCOMM Computer Communication Review* 40.4 (2010), pp. 375–386.
19. Ruben Torres et al. "Dissecting Video Server Selection Strategies in the YouTube CDN". In: *IEEE International Conference on Distributed Computing Systems (ICDCS)*. 2011.
20. Duc A. Tran, Khanh Nguyen, and Cuong Pham. "S-CLONE: Socially-Aware Data Replication for Social Networks". In: *Computer Networks* 56.7 (2012), pp. 2001–2013. ISSN: 1389–1286. DOI:10.1016/j.comnet.2012. 02.010. URL: http://www.sciencedirect.com/science/ article/pii/S1389128612000746.
21. Zhi Wang et al. "Enhancing Internet-scale Video Service Deployment Using Microblog-based Prediction". In: *IEEE Transactions on Parallel and Distributed Systems* 26.3 (Mar. 2015), pp. 775–785.
22. Zhi Wang et al. "Guiding Internet-Scale Video Service Deployment Using Microblog-based Prediction". In: *IEEE International Conference on Computer Communications (INFOCOM)*. 2012.

23. Zhi Wang et al. "Joint Social and Content Recommendation for User Generated Videos in Online Social Network". In: *IEEE Transactions on Multimedia* 15.3 (Apr. 2013), pp. 698–709.
24. Yu Wu et al. "Scaling Social Media Applications into Geo-Distributed Clouds". In: *IEEE International Conference on Distributed Computing Systems (ICDCS)*. 2012.
25. D. Xu et al. "Analysis of a CDN–P2P Hybrid Architecture for Cost-Effective Streaming Media Distribution". In: *Multimedia Systems* 11.4 (2006), pp. 383–399.
26. YouTube, http://www.youtube.com/yt/press/statistics.html.
27. W. Zhu et al. "Multimedia Cloud Computing". In: *IEEE Signal Processing Magazine* 28.3 (2011), pp. 59–69.

# Chapter 2
# Popularity of Social Videos

We begin by investigating the popularity of videos propagated through online social networks, including the popularity distribution and its evolution. Then, we investigate predictive models that capture social video popularity.

## 2.1 Social Video Popularity: Distribution and Evolution

To investigate the distribution of social video popularity, we used a real-world dataset from Renren (one of the largest online social networks in China), which contained log records for 1 week, during which 3 million users issued 11 million video shares and performed 87 million views [7]. We observed the following popularity characteristics.

### 2.1.1 Amplified Skewness in Popularity Distribution

The *Pareto* principle (also known as the 80–20 rule) is widely used to describe the skewness in distributions. For example, an analysis of YouTube shows that 10 % of the most popular videos account for 80 % of all user requests [2]. One may wonder whether social-network-based sharing results in a less skewed request distribution across the videos because all videos have an equal chance of becoming popular.

We first study the popularity of popular videos. In Fig. 2.1, we plotted the fraction of cumulative requests for videos against their popularity, ranked and normalized between 0 and 1. Figure 2.1 shows a counterintuitive result: just 0.4 % of the videos account for more than 80 % of the requests, while the remaining 99.6 % of the videos account for only 20 % of the requests. The reason for these results is that in online social networks, popular videos become even more popular because users

© The Author(s) 2016

Z. Wang et al., *Social Video Content Delivery*, SpringerBriefs in Electrical and Computer Engineering, DOI 10.1007/978-3-319-33652-7_2

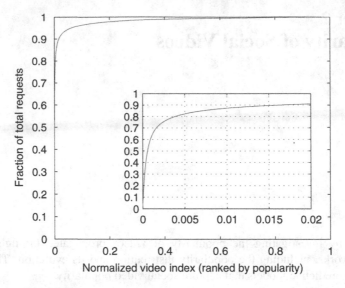

**Fig. 2.1** Skewness of requests across all videos

are more likely to recommend them to their friends. Less popular videos, however, fade out quickly in the small online social communities.

We further study the number of requests issued to these social videos in a time span of 1 day. As shown in Fig. 2.2, 80 % of the videos have fewer than four requests, and 90 % of the videos have fewer than ten requests per day. Considering the huge number of Renren users, this result confirms that social-network-based sharing will lead to a more skewed popularity distribution across the videos.

To further analyze the distribution of requests issued in different time spans, we examined videos that are shared on the first day. Because most users are more interested in newly updated videos, this analysis avoids possible bias from the age of the videos. We counted the cumulative requests for those videos within 1 day, 2 days, 1 week, and 1 month, respectively. Figure 2.3 shows the results. We observe that the skewness increases as the time window increases, and it converges after 1 week.

Based on the above analysis, we note that social-network-based sharing has affected the pattern of video popularity in existing video sharing systems. In particular, users' interests significantly converge on a few very popular videos. These videos are usually widely shared and recommended by many users, which helps them become even more popular. In contrast, the less popular videos fade out quickly in such an environment due to "social selection." We also observe that the popularity distribution is formed in a small time window of 1 week.

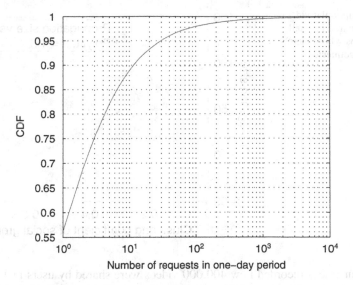

**Fig. 2.2** Distribution of request frequency in a time span of 1 day

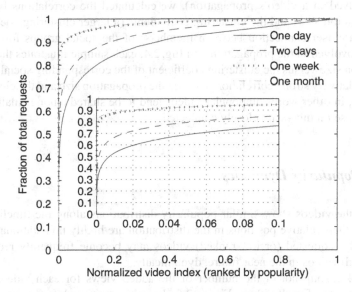

**Fig. 2.3** Skewness of requests across videos initially shared on the same day for different time spans

## 2.1.2   Sharing in Small Social Groups

Because the fraction of unpopular social videos is so large, we investigated the types of social groups in which these unpopular videos propagate, using traces from

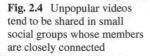

**Fig. 2.4** Unpopular videos tend to be shared in small social groups whose members are closely connected

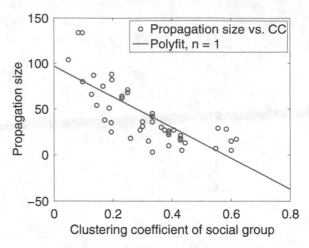

Weibo. Our traces recorded how 400,000 videos were shared by users in 1 month. By randomly sampling 50 videos with different propagation sizes (the number of users involved in a video's propagation), we calculated the correlations between the propagation sizes and the clustering coefficients (a larger clustering coefficient indicates a closer connection between members) of the social groups formed by the users involved in the propagations. In Fig. 2.4, each sample illustrates the video propagation size versus the clustering coefficient of the corresponding social group. There is relatively strong correlation between the propagation size and the clustering coefficient; in other words, unpopular videos tend to be shared among small social groups whose members are closely connected.

### 2.1.3 Popularity Dynamics

Although the videos show similar popularity distributions along the timeline, we found that their relative positions in the distribution are highly non-stationary, i.e., some rarely requested (or low-ranked) videos may become frequently requested (top-ranked) videos in the near future, dynamically.

We took a snapshot of the number of the added views for each video, every 500,000 requests for all videos. Figure 2.5 shows the scatter plots for the number of new views received by a video at snapshots 2, 3, and 4 against the number at snapshot 1. We calculate both the Pearson correlation coefficient ($\rho_p$) [5] and Spearman's rank correlation coefficient ($\rho_s$) [4][1] between the number of new views

---

[1] $\rho_p$ has been widely used to measure the strength of linear dependence between two variables, and $\rho_s$ assesses how well the relationship between two variables can be described using a monotonic function. The ranges of both $\rho_p$ and $\rho_s$ are from $-1$ to $1$, where a value greater than 0 indicates a

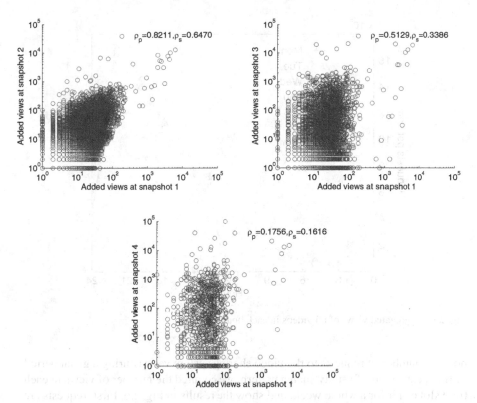

**Fig. 2.5** The number of new views at snapshots 2, 3, and 4 versus that at snapshot 1

at different snapshots and the original number of views at snapshot 1. This figure demonstrates the change in the number of views between two snapshots. Overall, there is substantial non-stationarity in the popularity of individual videos: Although the new views for two adjacent snapshots show a weak correlation, non-adjacent snapshots show that the correlation declines quickly as the distance between any two snapshots increases.

## 2.1.4 Popularity Evolution

We measured the evolution of video popularity in online social networks. Similar to most streaming systems, our measurement shows that video requests are not distributed uniformly but exhibit a diurnal pattern. The diurnal access pattern defines

---

positive correlation and a value less than 0 indicates a negative correlation. A value of 0.8 or more is usually considered strong positive correlation [2].

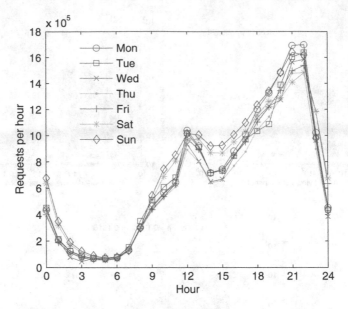

**Fig. 2.6** Aggregated views of all videos in each hour in 1 week

how the number of requests to the video sharing system varies during a given period of time, e.g., a day. To study such a pattern, we counted the number of views in each time slot of 1 h for a whole week, and show the results in Fig. 2.6. First, requests are slightly elevated over weekends. Second, there are local peaks around lunchtime, especially on weekdays. Third, although it is common for the fewest requests to appear in the early morning (at approximately 6 am) and for the most requests to appear at approximately 10 pm, the large gap between the peak values and the lowest values (by 20 times) exceeded our expectations.

The diurnal social video request pattern is important for capturing the bursts of resource consumption within a given time period. Due to the diurnal request pattern, the inter-arrival times within a given day do not simply follow an exponential distribution. Instead, it can be better modeled as a non-homogeneous Poisson process [4]. The request number in a given time slot can be computed based on the diurnal pattern.

We further examined the popularity evolutionary patterns of videos with varying popularity. From a 3-month period, we randomly selected 1000 high-popularity videos with more than 10,000 total requests per video, 1000 medium-popularity videos with 400–600 requests per video, and 1000 low-popularity videos with ≤10 requests per video. Figure 2.7 shows the popularity evolution of the three groups of videos. As shown, the low-popularity videos attract only a few users even on the first day when they are shared. Subsequently, their popularity decreases quickly to a near-zero level, showing that the average lifetime of a low-popularity video is less than 1 day, suggesting that users quickly

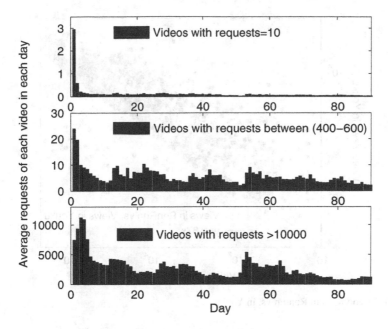

**Fig. 2.7** Popularity evolution of videos with varying popularity

lose interest in these videos in the online social network. The medium-popularity videos, on the other hand, show a very different evolutional pattern over time. Although they also achieve their request peaks during the first day, their popularity decreases much more slowly than the low-popularity videos. In contrast, the peak value for high-popularity videos generally arrives only after 2 or 3 days, and these videos remain popular at a relatively high level for a long time.

## 2.1.5 Popularity Comparison

We explored whether videos have similar popularity in online social networks and traditional video sharing services. Among all video sharing sites that ever share videos in Renren, Youku alone accounts for nearly 80 % of all shared video views, while the top five video sites together account for nearly 95 % of all the requests. Figure 2.8 shows the video views in Renren compared with those in Youku.

The sites exhibit a relatively close popularity relationship, which indicates that the content itself plays an important role in a video's popularity. The fitted curve suggests that approximately 37 % (1/2.67) of the video views on Youku come from Renren. Note that we considered only the statistics of those Youku videos that were ever shared in Renren. The ratio of videos from Youku that were ever shared in Renren was approximately 11 % in March 2011. The ratio had increased to 15 %

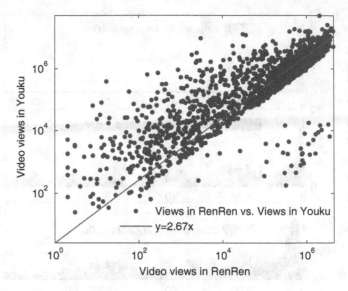

**Fig. 2.8** Video views in Renren vs. in Youku

when we measured it again in October, 2011. Considering the benefits of this interaction for both online social networks and traditional video sharing services, we believe that this ratio of videos shared in online social networks will continue to increase steadily in the future.

To analyze the correlation between the number of video views in Renren and Youku with more specificity, we use the Pearson correlation coefficient ($\rho_p$) and Spearman's rank correlation coefficient ($\rho_s$). The value of $\rho_s$ is 0.84, indicating a relatively high positive correlation and confirming the result in Fig. 2.8. As a comparison, $\rho_p$ is 0.69, which is much smaller than $\rho_s$ and indicates that the popularity of videos in Renren and Youku does not have a good linear correlation relationship.

To understand the type of videos that are popular in Renren, we show the correlation coefficients between the video views in Renren and three other statistics in Table 2.1, including likes, dislikes, and comments. Our observations are as follows. (1) The number of views has the highest correlation with traditional video sharing services. Because a video is generally first uploaded to a traditional video sharing service (e.g., YouTube) and then discovered and shared by users in online social networks (usually after the video becomes popular), a video's popularity in a traditional video sharing service can be used as an important predictor of its popularity when it is first shared in online social networks. (2) It is surprising that there is little correlation between the number of likes/comments and the number of views.

**Table 2.1** Correlation between video views and other statistics in Renren and Youku

| Correlation | Views | Likes | Dislikes | Comments |
|---|---|---|---|---|
| $\rho_p$ | 0.6937 | 0.2780 | 0.0188 | 0.2203 |
| $\rho_s$ | 0.8493 | 0.6801 | 0.6186 | 0.6189 |

## 2.2 Social Popularity Prediction

The above observations suggest that content delivery mechanisms must be substantially revised, and that social-aware factors (e.g., the number of previous re-sharers and the age of the video in an online social network) will play important roles [1]. Meanwhile, the popularity prediction has to jointly consider both accuracy and timeliness of the prediction results.

### 2.2.1 Social Popularity Prediction Framework

To capture the popularity of social video content, we investigated a systematic methodology and an associated online algorithm for forecasting the popularity of videos promoted by social media. Our social-forecast algorithm is able to make predictions about the future popularity of videos while jointly considering both the accuracy and the timeliness of the prediction. We consider the unique situational conditions that affect videos propagated in social media, and we demonstrate how this contextual information can be incorporated to improve the accuracy of the forecasts.

We considered a generic Web 2.0 information sharing system in which videos are shared by users through social media. The framework is illustrated in Fig. 2.9. We assigned each video an index $k \in \{1, 2, \ldots, K\}$ according to the absolute time $t_{init}^k$ when it is first posted.[2]

After a video is first posted, it will propagate through social media for some time duration. We assume a discrete time model in which a period can be minutes, hours, days, or any suitable time duration. A video has an age of $n \in \{1, 2, \ldots\}$ periods if it has been propagated through social media for $n$ periods. In each period, the video is further shared and viewed by users depending on the sharing and viewing status of the previous period. The propagation characteristics of video $k$ up to age $n$ are captured by a $d_n$-dimensional vector $\mathbf{x}_n^k \in \mathscr{X}_n$, which contains popularity information such as the total number of views, as well as other contextual information such as the characteristics of the social network over which the video was propagated. We keep $\mathbf{x}_n^k$ in an abstract form and call it the *context (and situational) information* at age $n$.

---

[2]It is easy to assign unique identifiers for multiple videos that are generated/initiated at the same time.

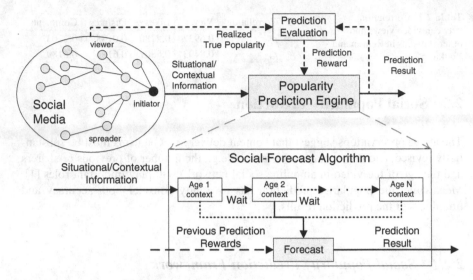

**Fig. 2.9** Social-aware popularity prediction framework

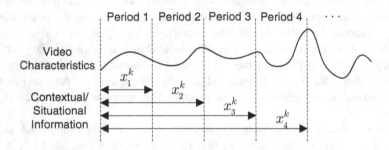

**Fig. 2.10** An illustration of contextual information taking the history characteristics into account

Several points are worth noting regarding the contextual information.

First, the context space $\mathcal{X}_n$ can be different at different ages $n$. In particular, $\mathbf{x}_n^k$ can include all the history information for video $k$'s propagation characteristics up to age $n$, i.e., $\mathbf{x}_n^k$ includes all the information for $\mathbf{x}_m^k$, $\forall m < n$, as illustrated in Fig. 2.10.

Second, the contextual/situational information $\mathbf{x}_n^k$ can be taken from a large space, e.g., a finite space with a large number of values or even an infinite space. For example, some dimensions of $\mathbf{x}_n^k$ take values from a continuous value space and $\mathbf{x}_n^k$ may include all the past propagation characteristics (e.g., $\mathbf{x}_m^k \in \mathbf{x}_n^k$, $\forall m < n$).

Third, at age $n$, $\mathbf{x}_m^k$, $\forall m > n$ are not yet revealed because they represent future situational and contextual information that is yet to be realized. Hence, given the contextual information represented by $\mathbf{x}_n^k$ at age $n$, the future contextual information $\mathbf{x}_m^k$, $\forall m > n$ are random variables.

We are interested in predicting the future popularity status of videos by the end of a pre-determined age $N$, and the goal is to make that prediction as soon as possible.

The choice of $N$ depends on the specific requirements of the content provider, the advertiser, and the web hosts. We will treat $N$ as given, because the video sharing events have daily and weekly patterns, and the active lifespans of most shared videos through social media are quite limited [3]. Thus, the contextual information for video $k$ during its lifetime of $N$ periods is collected in $\mathbf{x}^k = (\mathbf{x}_1^k, \mathbf{x}_2^k, \ldots, \mathbf{x}_N^k)$. For expositional simplicity, we also define $\mathbf{x}_{n+} = (\mathbf{x}_{n+1}, \ldots, x_N)$, $\mathbf{x}_{n-} = (\mathbf{x}_1, \ldots, x_{n-1})$ and $\mathbf{x}_{-n} = (\mathbf{x}_{n-}, \mathbf{x}_{n+})$.

Let $\mathscr{S}$ denote the popularity status space, which is assumed to be finite. For instance, $\mathscr{S}$ can be either a binary space {popular, unpopular} or a more refined space containing multiple levels of popularity such as {low popularity, medium popularity, high popularity} or any such refinement. Let $s^k$ denote the popularity status of video $k$ by the end of age $N$. Because $s^k$ is realized only at the end of $N$ periods, it is a random variable at all previous ages. However, the conditional distributions of $s^k$ will vary at different ages because those distributions are conditioned on different contextual information. In many scenarios, the conditional distribution at a higher age $n$ is more informative for predicting future popularity because more contextual information is available. Nevertheless, our model does not require this assumption to hold.

### 2.2.2  Prediction Reward

For each video $k$, at each age $n = 1, \ldots, N$, we can make a prediction decision $a_n^k \in \mathscr{S} \cup \{\text{wait}\}$. If $a_n^k \in \mathscr{S}$, we predict $a_n^k$ as the popularity status by age $N$. If $a_n^k = \text{wait}$, we choose to wait for the next period's contextual information to decide, i.e., either predict a future popularity status or wait again. When the prediction is used to make a one-shot decision, introducing a "wait" option is of significant importance to allow a trade-off between accuracy and timeliness. For each video $k$, at the end of age $N$, given the decision action vector $a^k$, we define the *age-dependent reward* $r_n^k$ at age $n$ as follows:

$$r_n^k = \begin{cases} U(a_n^k, s^k, n), & \text{if } a_n^k \in \mathscr{S} \\ r_{n+1}^k, & \text{if } a_n^k = \text{wait} \end{cases} \qquad (2.1)$$

where $U(a_n^k, s^k, n)$ is a reward function that depends on the accuracy of the prediction (determined by $a_n^k$ and the realized true popularity status $s^k$) and the timeliness of the prediction (determined by the age $n$ when the prediction is made).

The specific form of $U(a_n^k, s^k, n)$ depends on how the reward is derived according to the popularity prediction based on various economic and technological factors. For instance, the reward can be set as the ad revenue derived from placing appropriate ads with potentially popular videos, or it can be the cost of adequately planning computation, storage, and bandwidth resources to ensure the robust operation of the video streaming services. A reward function can be in the form of

$U(a_n^k, s^k, n) = \theta(a_n^k, s^k) + \lambda \psi(n)$, where $\theta(a_n^k, s^k)$ measures the prediction accuracy, $\psi(n)$ accounts for the prediction timeliness, and $\lambda > 0$ is a trade-off parameter that controls the relative importance of accuracy and timeliness.

Let $n^*$ be the first age at which the action is not wait, i.e., the first time a forecast is issued. The *overall prediction reward* is defined as the $r^k = r_{n^*}^k$. According to the equation reward, when the action is wait at age $n$, the reward is the same as that at age $n+1$. Thus $r_1^k = r_2^k = \ldots = r_{n^*}^k$. This suggests that the overall prediction reward is the same as the age-dependent reward at age 1, i.e., $r^k = r_1^k$. For age $n > n^*$, the action $a_n^n$ and the age-dependent reward $r_n^k$ do not affect the realized overall prediction result because a prediction has already been made. However, we still select actions and compute the age-dependent reward because it helps in learning the best action and the best reward for this age $n$—which in turn will help decide whether we should wait at an early age. Figure 2.11 provides an illustration on how the actions at different ages determine the overall prediction reward.

*Remarks.* The prediction action itself does not generate rewards. It is the action taken using the prediction results that is rewarding. In many scenarios, this action can only be taken once and cannot be altered afterwards. This motivated the formulation of the above overall reward function in which the overall prediction reward is determined by the first non-wait action. Nevertheless, our framework could easily be extended to account for more general overall reward functions which may depend on all non-wait actions. For instance, the action could be revised when a more accurate later prediction is made. In this case, the reward function $U(a_n^k, s^k, n)$ in reward would depend not only on the current prediction action $a_n^k \in \mathscr{S}$ but also on all non-wait actions after age $n$. We will use the reward function shown in reward because of its simplicity for the exposition—but our analysis also holds for general reward functions.

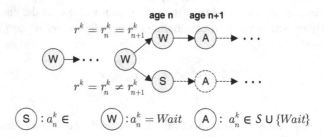

**Fig. 2.11** An illustration of the multi-stage decision making. If the first $n$ actions are wait, then $r_n^k = r_{n+1}^k$ which depends on later actions. If the age-$n$ action is not wait, then $r_n^k \neq r_{n+1}^k$, and $r^k$ does not depend on later actions. However, we can still determine the reward of taking action at age $n + 1$ just as if all actions before age $n + 1$ were wait

### 2.2.3 Prediction Policy

We focus on prediction policies that depend on current contextual information. Let $\pi_n : \mathscr{X}_n \to \mathscr{S} \cup \{\text{wait}\}$ denote the prediction policy for a video of age $n$ and $\pi = (\pi_1, \ldots, \pi_N)$ be the complete prediction policy. Hence, a prediction policy $\pi$ contains the actions for all possible contextual information at all ages. For expositional simplicity, we also define $\pi_{n+} = (\pi_{n+1}, \ldots, \pi_N)$ as the policy vector for ages greater than $n$, $\pi_{n-} = (\pi_1, \ldots, \pi_{n-1})$ as the policy vector for ages smaller than $n$ and $\pi_{-n} = (\pi_{n-}, \pi_{n+})$. For a video with contextual information $\mathbf{x}^k$, the prediction policy $\pi$ determines the prediction action at each age and hence the overall prediction reward, denoted by $r(\mathbf{x}|\pi)$, as well as the age-dependent rewards $r_n(\mathbf{x}|\pi), \forall n = 1, \ldots, N$. Let $f(\mathbf{x})$ be the probability distribution function of the video's contextual information, which also gives information about the popularity evaluation patterns. The expected prediction reward of a policy $\pi$ is, therefore,

$$V(\pi) = \int_{\mathbf{x} \in \mathscr{X}} r(\mathbf{x}|\pi) f(\mathbf{x}) d\mathbf{x}. \tag{2.2}$$

Note that the age-$n$ policy $\pi_n$ will use the contextual information $\mathbf{x}_n$ rather than $\mathbf{x}$ to make predictions because $\mathbf{x}_{n+}$ has not yet been realized at age $n$.

Our objective is to determine the optimal policy $\pi^{opt}$ that maximizes the expected prediction reward, i.e., $\pi^{opt} = \arg\max_{\pi} V(\pi)$. This problem is referred to as a complete information problem if $f(\mathbf{x})$ is known, and incomplete information problem otherwise.

### 2.2.4 Prediction Using Complete Information

We consider the optimal policy design problem with the complete information of the context distribution $f(\mathbf{x})$ and compute the optimal policy $\pi^{opt}$. Even when having the complete information, determining the optimal prediction policy faces great challenges: First, the prediction reward depends on all decision actions at all ages; Second, when making the decision at age $n$, the actions for ages larger than $n$ are not known since the corresponding context information has not been realized yet.

Given policies $\pi_{-n}$, we define the expected reward when taking action $a_n$ for $\mathbf{x}_n$ as follows:

$$\mu_n(\mathbf{x}'_n | \pi_{-n}, a_n) = \int_{\mathbf{x}} I_{\mathbf{x}_n = \mathbf{x}'_n} r_n(\mathbf{x}|\pi_{-n}, a_n) f(\mathbf{x}) d\mathbf{x} \tag{2.3}$$

where $I_{\mathbf{x}_n = \mathbf{x}'_n}$ is an indicator function which takes value 1 when the age-$n$ context information is $\mathbf{x}'_n$ and value 0 otherwise. The optimal $\pi^*(\pi_{-n})$ given $\pi_{-n}$ thus can be determined by

$$\pi_n^*(\mathbf{x}_n|\pi_{-n}) = \arg\max_a \mu(\mathbf{x}_n|\pi_{-n}, a), \forall \mathbf{x}_n. \tag{2.4}$$

Equation (2.4) defines a best response function from a policy to a new policy $F$ : $\Pi \to \Pi$ where $\Pi$ is the space of all policies. In order to compute the optimal policy $\pi^{opt}$, we iteratively use the best response function in (2.4) using the output policy computed in the previous iteration as the input for the new iteration. Note that a computation iteration is different from a time period, which is used to describe the time unit of the discrete time model of the video propagation. A period can be a minute, an hour, or any suitable time duration. In each period, the sharing and viewing statistics of a specific video may change. "Iteration" is used for the (offline) computation method for the optimal policy (which prescribes actions for *all* possible context information in *all* periods). Given the complete statistical information (i.e., the video propagation characteristics distribution $f(\mathbf{x})$) of videos, a new policy is computed using best response update in each iteration.

We prove the convergence and optimality of this best response update as follows.

**Lemma 2.1.** $\pi_n^*(\mathbf{x}_n|\pi_{-n})$ *is independent of* $\pi_m, \forall m < n$, *i.e.* $\pi_n^*(\mathbf{x}_n|\pi_{-n}) = \pi_n^*(\mathbf{x}_n|\pi_{n}+)$.

*Proof.* By the definition of age-dependent reward, the prediction actions before age $n$ do not affect the age-$n$ reward. Hence, the optimal policy depends only on the actions after age $n$.

Lemma 2.1 shows that the optimal policy $\pi_n$ at age $n$ is fully determined by the policies for ages larger than $n$ but does not depend on the policies for ages less than $n$. Using this result, we can show the best response algorithm converges to the optimal policy within a finite number of computation iterations.

**Theorem 2.1.** *Starting with any initial policy* $\pi^0$, *the best response update converges to a unique point* $\pi^*$ *in* $N$ *computation iterations. Moreover,* $\pi^* = \pi^{opt}$.

*Proof.* Given the context distribution $f(\mathbf{x})$ which also implies the popularity evolution, the optimal age-$N$ policy can be determined in the first iteration. Since we break ties deterministically when rewards are the same, the policy is unique. Given this, in the second iteration, the optimal age-$(N-1)$ policy can be determined according to (2.4) and is also unique. By induction, the best response update determines the unique optimal age-$n$ policy after $N+1-n$ iterations. Therefore, the complete policy is found in $N$ iterations and this policy maximizes the overall prediction reward.

Theorem 2.1 proves that we can compute the optimal prediction policy using a simple iterative algorithm as long as we have complete knowledge of the popularity evolution distribution. In practice, this information is unknown and extremely difficult to obtain, if not possible. One way to estimate this information is based on a training set. Since the context space is usually very large (which usually involves infinite number of values), a very large volume of training set is required to obtain a reasonably good estimation. Moreover, existing training sets may be biased and outdated as social media evolves. Hence, prediction policies developed using existing training sets may be highly inefficient [6]. In [7], we also developed learning algorithms to learn the optimal policy in an online fashion, requiring no initial knowledge of the popularity evolution patterns.

To summarize, we formulate the online popularity prediction as a multi-stage sequential decision and online learning problem. Our solution makes multi-level popularity prediction in an online fashion and requires no a priori training phase or dataset. It exploits the dynamically changing and evolving video propagation patterns through social media to maximize the prediction reward. The algorithm is easily tunable to enable trade-offs between the accuracy and timeliness of the forecasts as required by various applications, entities, and/or deployment scenarios.

## 2.3  Summary

In this chapter, we studied the characteristics of social video popularity and its prediction. We presented the changes of popularity distribution of social video content, including the amplified skewness and the dynamical popularity evolution. Based on these observations, we next investigated social video popularity prediction: an age-dependent model has been proposed to infer social video popularity using both content and context information.

## References

1. Youmna Borghol et al. "The Untold Story of the Clones: Content-agnostic Factors That Impact YouTube Video Popularity". In: *ACM SIGKDD Conference on Knowledge Discovery and Data Mining (KDD)*. 2012.
2. M. Cha et al. "I Tube, You Tube, Everybody Tubes: Analyzing the World's Largest User Generated Content Video System". In: *ACM SIGCOMM*. 2007.
3. Haitao Li, Xu Cheng, and Jiangchuan Liu. "Understanding Video Sharing Propagation in Social Networks: Measurement and Analysis". In: *ACM Transactions on Multimedia Computing, Communications, and Applications (TOMM)* 10.4 (2014), p. 33.
4. J. S. Maritz. *Distribution-free Statistical Methods*. 1995.
5. J. L. Rodgers and W. A. Nicewander. *Thirteen Ways to Look at the Correlation Coefficient*. The American Statistician, 1988.
6. Peter Sollich and David Barber. "Online learning from finite training sets and robustness to input bias". In: *Neural computation* 10.8 (1998), pp. 2201–2217.
7. Jie Xu et al. "Forecasting Popularity of Videos using Social Media". In: *arXiv preprint arXiv:1403.5603* (2014).

# Chapter 3
# Dynamical Social Video Propagation

The popularity of a shared video reflects its macroscopic aggregated views. In this chapter, we take a closer look at the social propagation process for videos, which determines how individual videos reach different users in online social networks.

## 3.1 Dynamic Social Video Propagation

The generation and re-sharing of a social video typically form a propagation tree rooted at the user who generates the video or initiates the sharing (referred to as the *initiator* or *root*). We refer to the users who re-share the video as *spreaders* and users who receive the shared video as *viewers* (or *receivers*). A video's popularity can then be calculated as the sum of its spreaders and receivers.

In addition to the normal nodes that view and share videos, there are two types of nodes that are worth highlighting. First, there are super spreaders during propagation, who are followed by many viewers. The super spreaders—especially those who appear early in the propagation stage—generally play an important role in the further "explosion" of the propagation (i.e., attracting many viewers). Second, there are free riders during propagation, who do not share videos at all and comprise a large portion of the viewers. Free riders only consume videos shared by others.

An epidemic model describes the spread of a contaminative disease through a population [2]. One classical epidemic model is the Susceptible-Infectious-Recovered (SIR) model, as follows. Initially, a user shares a video from an external video sharing site, and this initiator becomes *infectious*. All other users in the social network are *safe* except the friends of the initiator. The shared video appears in the news feed of the initiator's friends and, thus, they become *susceptible*. Over time, these friends gradually log into the social network and decide whether to watch the video (*infected*) or not (*immune*). The *infected* users will decide whether to

© The Author(s) 2016
Z. Wang et al., *Social Video Content Delivery*, SpringerBriefs in Electrical and Computer Engineering, DOI 10.1007/978-3-319-33652-7_3

share after watching the video. They become *recovered* if they choose not to share, and *infectious* if they choose to share. Again, these infectious users will make their friends who are in the safe stage become susceptible.

Based on the propagation model, the parameters can be trained using propagation data extracted from real-world logs to identify the connection between video access and social activities to improve content delivery strategies. Large-scale measurement studies have discovered interesting locality patterns in the propagation structures [6].

### 3.1.1  Social Locality

The generation and re-share of a video in an online social network forms a propagation tree rooted at the user who first posts the video. Any user who re-shares that video becomes a new branch node in the propagation tree. Figure 3.1 shows the propagation sizes of videos in five different categories. Each sample illustrates the number of propagation trees (with the same propagation size) versus the size of these propagation trees. The size of most propagation trees is very small, e.g., the size of over 90 % of the propagation trees is smaller than 100.

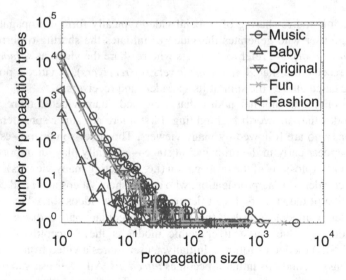

**Fig. 3.1** Number of propagation trees versus the propagation size

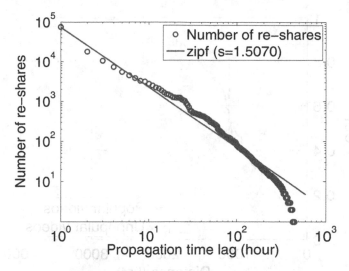

**Fig. 3.2** Number of re-shares versus the time lag

### 3.1.2 Temporal Locality

In an online social network, users are more likely to re-share new video content, i.e., videos that have been recently imported or re-shared. Figure 3.2 illustrates the number of re-shares of a video in a timeslot (1 h) versus the time lag since the propagation started. As shown, most of the re-shares happen in the most recent hours, and the re-share number against the time lag follows a zipf-like distribution with a shape parameter of $s = 1.5070$. More than 95 % of the re-shares happen within the first 24 h, indicating that users' behaviors in social video sharing are highly clustered around the time point when the video is first shared.

### 3.1.3 Geographical Locality

A large fraction of online videos are shared between users who are geographically close to each other. In Fig. 3.3, we plot the CDF of distances between users who join in the social propagation of the same video within an online social network. The various curves show videos with different popularities as follows: (1) Popular videos are those whose popularity is in the top 2 %; (2) Unpopular videos are those whose popularity is in the bottom 30 %. In contrast to traditional video consumption, unpopular social videos tend to be shared in local regions (e.g., the same city), where users are located close to each other. For example, approximately 40 % of the distances between users sharing the same unpopular videos are close to 0 km.

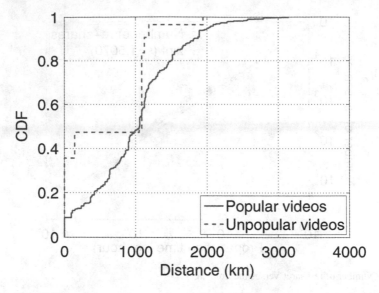

**Fig. 3.3** CDF of distances between users who are sharing in the same propagation of social videos

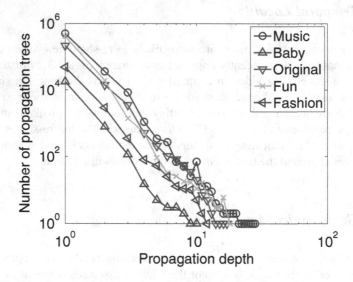

**Fig. 3.4** Number of propagation trees versus the propagation depth

Next, we studied the propagation depth, which is defined as the average number of social hops between users in the propagation tree and the root user. Figure 3.4 illustrates the propagation depth of videos in the same five categories. Each sample represents the number of propagation trees (with the same propagation depth) versus their propagation depth. In most of the propagation trees, the depth does not exceed

10, i.e., users who re-share the same video are socially close to the root user, with only a small number of social hops between them.

The limited propagation size and propagation depth indicate that in each propagation tree, only users within a *limited social range* will be reached by the video. This observation motivated us to design a peer-assisted replication feature so that users who are both socially and geographically close to each other help distribute videos among themselves effectively. The details will be presented in Chap. 4.

We further studied the propagation structure of social videos and identified a series of representative structures [1]. We defined a *branching factor* as the number of viewers that directly follow a spreader and a *share rate* as the ratio of the viewers who re-share the video after watching it. An interesting observation here is that the branching factor and share rate are depth-independent: they are merely correlated to the users' distance (the number of social hops) from the root. As such, the branching factor and share rate can be set to the same values for all spreaders and viewers, regardless of the social distance to the root.

## 3.2 Modeling Social Video Content Propagation

In this section, we propose an extended epidemic model to capture video propagation in online social networks. First, we describe the classical SIR model and extend it to our $S^2I^3R$ model. Then, we validate it based on real-world log data. Finally, using this $S^2I^3R$ model, we analyze an interesting observation from the measurements.

### 3.2.1 A $S^2I^3R$ Propagation Model

The classical epidemic model, SIR , which considers a fixed population with three compartments [2]: *Susceptible* (S), *Infectious* (I), and *Recovered* (R). The initial letters also represent the number of people in each compartment at a particular time $t$, that is, at any time $t$, $S(t)$ is the number of individuals not yet infected with the disease—those susceptible to the disease; $I(t)$ is the number of individuals who have been infected with the disease and are capable of spreading it to those in the susceptible category; and $R(t)$ is the compartment used for those individuals who have been infected and then recovered from the disease. People in this category are neither able to be infected again nor to transmit the infection to others.

In the SIR model, we have the following ordinary differential equations:

$$\begin{cases} \dfrac{dS(t)}{dt} = -\beta \cdot S(t) \cdot I(t) & (3.1) \\[2ex] \dfrac{dI(t)}{dt} = \beta \cdot S(t) \cdot I(t) - \gamma \cdot I(t) & (3.2) \\[2ex] \dfrac{dR(t)}{dt} = \gamma \cdot I(t) & (3.3) \end{cases}$$

where parameter $\beta$ is the infection rate of the disease, and parameter $\gamma$ represents the recovery rate.

No direct mapping exists from the classical SIR model for video propagation in an online social network; therefore, new compartments and new derivative equations are required. The major compartments in a video propagation context are as follows:

- Safe ($S_1$) represents individuals who are far away from sharers. Initially, all users are Safe except the friends of the initiator;
- Susceptible ($S_2$) represents the individuals who have a chance to see the shared video. If an individual shares a video, the shared video will appear in that person's friends' news feeds, and any friends formerly in the Safe stage become Susceptible;
- Infected ($I_1$) represents the individuals who are watching the video. Note that individuals at this stage still cannot infect others;
- Immune ($I_2$) denotes the individuals who choose not to watch the video;
- Infectious ($I_3$) denotes the individuals who choose to share the video after watching it. Only individuals who are at the Infectious stage can infect other individuals;
- Recovered ($R$) denotes the individuals who have watched the video, but choose not to share it.

In the classical SIR model, the transition is time-dependent; in other words, at any time, there is a chance that a particular stage transits to the next stage. In contrast, for video sharing propagation in online social networks, the transition between stages depends on decisions made at a certain time, e.g., the user chooses to watch or not watch and, then, to share or not share. Therefore, we investigated an extended epidemic model to estimate the video propagation [4], by introducing two temporary decision stages in $S^2I^3R$: D1 and D2. The user makes the watching decision at stage D1 and makes the sharing decision at stage D2.

The enhanced Safe-Susceptible-Infected-Immune-Infectious-Recovered ($S^2I^3R$) model is illustrated in Fig. 3.5. For a particular video object, the propagation process

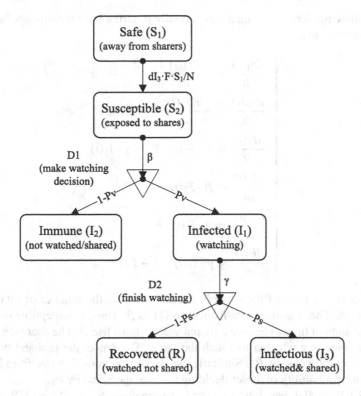

**Fig. 3.5** $S^2I^3R$ model

moves through these stages: Initially, some user (the initiator) shares this video from an external video sharing site and thereby becomes infectious. All other users in the social network are safe except the friends of the initiator. The shared video appears in the news feed of the initiator's friends and thus they become susceptible. Over time, these friends gradually log into the social network and decide whether to watch the video (infected) or not (immune). The infected users will usually decide whether share after watching the video. They become recovered if they choose not to share, or infectious if they choose to share. Again, these infectious users cause their friends who are in the safe stage to become susceptible. Note that the case of "not watched but shared" is not considered in the $S^2I^3R$ model. That is because this case accounts for only a small portion (e.g., less than 5 %) among all "share" cases. Moreover, omitting this case lets us simplify the model and focus on the more important parameters.

The following derivative equations formally describe the relationships between those compartments:

$$
\begin{cases}
\dfrac{dS_1(t)}{dt} = -\dfrac{dI_3(t)}{dt} \cdot F \cdot \dfrac{S_1(t)}{N} & \text{(3.4)} \\[2ex]
\dfrac{dS_2(t)}{dt} = -\dfrac{dS_1(t)}{dt} - \beta \cdot S_2(t) & \text{(3.5)} \\[2ex]
\dfrac{dI_1(t)}{dt} = \beta \cdot S_2(t) \cdot P_v - \gamma \cdot I_1(t) & \text{(3.6)} \\[2ex]
\dfrac{dI_2(t)}{dt} = \beta \cdot S_2(t) \cdot (1 - P_v) & \text{(3.7)} \\[2ex]
\dfrac{dI_3(t)}{dt} = \gamma \cdot I_1(t) \cdot P_s & \text{(3.8)} \\[2ex]
\dfrac{dR(t)}{dt} = \gamma \cdot I_1(t) \cdot (1 - P_s) & \text{(3.9)}
\end{cases}
$$

where $F$ is the number of the sharer's friends and $N$ is the number of total users in the system. The transition rate from S to D1 is $\beta$. Thus, a susceptible user will spend $1/\beta$ units of time to receive a shared video from a friend. The user then makes a decision regarding whether to watch the video. We denote the probability of the user watching the video as $P_v$. Similarly, we denote the transition rate from I to D2 by $\gamma$ and the probability of a user deciding to share the video by $P_s$.

The $S^2I^3R$ model has four important parameters: $\beta$, $\gamma$, $P_v$, and $P_s$. These parameters can be investigated using real-world log data. Specifically, for the Renren system, the cumulative distribution function of $1/\beta$, the time span from share to watch, is well fitted by a combined Weibull and a Generalized Pareto distribution

$$
f_{k,\lambda,\mu,\sigma,\xi}(x) = \begin{cases}
1 - e^{-(x/\lambda)^k} & x \le 2100 \\[2ex]
1 - \left(1 + \xi \cdot \dfrac{x-\mu}{\sigma}\right)^{-\frac{1}{\xi}} & x > 2100
\end{cases}
$$

with parameters ($k = 0.392$, $\lambda = 1945$, $\mu = -2654$, $\sigma = 6315$, $\xi = -0.669$). The cumulative distribution function of $1/\gamma$, the time span between watching a video and sharing it, is well fitted by two combined Weibull distributions

$$
f_{k_1,\lambda_1,k_2,\lambda_2}(x) = \begin{cases}
1 - e^{-(x/\lambda_1)^{k_1}} & x \le 5 \\[2ex]
1 - e^{-(x/\lambda_2)^{k_2}} & x > 5
\end{cases}
$$

with parameters ($k_1 = 1.168$, $\lambda_1 = 3.591$, $k_2 = 0.497$, $\lambda_2 = 2.129$)

The cumulative distribution functions of both $P_v$ and $P_s$ follow a Generalized Pareto Distribution

$$f_{\mu,\sigma,\xi}(x) = 1 - \left(1 + \xi \cdot \frac{x - \mu}{\sigma}\right)^{-\frac{1}{\xi}}$$

with parameters ($\mu = -0.004$, $\sigma = 0.182$, $\xi = -0.215$) and ($\mu = -0.227$, $\sigma = 0.305$, $\xi = -0.048$), respectively.

### 3.2.2 Model Validation

We have simulated the $S^2I^3R$ model multiple times to validate its accuracy. We generate 10,000 users participating in 100 video sharing propagations for 8640 min (6 days). Specifically, we simulate the propagation of one video each time and run it 100 times. For each video propagation, at each minute the simulator checks and updates the state for each of the 10,000 users according to the derivation Eqs. (3.4)–(3.9). The simulation runs 8640 cycles for each video propagation. The work [3] provided a distribution of the number of friends in Renren that we use in our simulation.

We extract a series of statistics such as the number of received, watched, and shared videos for each user, the time span from share to watch, and the time span from watch to share. We examine these statistics with the real dataset; specifically, we compute $R^2$, the *coefficient of determination*[1] of the generated data and the real data. We list the goodness of fit values, as well as the statistical fitting model names and the corresponding $R^2$ values from the simulation in Table 3.1. The high values of $R^2$ (above 0.99) indicate that our model accurately characterizes the users' behaviors in video propagation.

We next investigate the evolutionary patterns of the number of users at each stage along the timeline. We again generate 10,000 users participating in 100 video sharing propagations for 8640 min (6 days). We run the model 100 times using

**Table 3.1** Validation of $S^2I^3R$ model

|  | Fitting model | $R^2$ of fitting model | $R^2$ of simulation |
|---|---|---|---|
| Reception rate | GPD | 0.9978 | 0.9952 |
| Share rate | GPD | 0.9959 | 0.9540 |
| Time to watch | Weibull + GPD | 0.9991 | 0.9348 |
| Time to share | 2 Weibulls | 0.9989 | 0.9813 |

[1] The coefficient of determination $R^2$ is a goodness of fit statistic describing how well a variable fits a set of observations, defined as $1 - \frac{\sum_i (y_i - f_i)^2}{\sum_i (y_i - \bar{y})^2}$, where $f$ are generated data or modeled values, $y$ are the real data, and $\bar{y}$ is the mean of the real data.

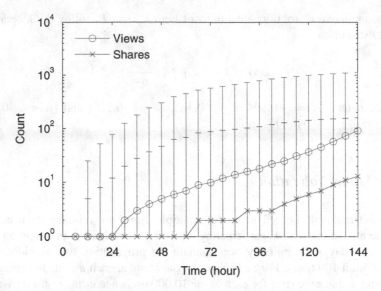

**Fig. 3.6** Cumulated video views ($I_3 + R$) and video shares ($I_3$) along time

the same system settings (including $\beta$, $\gamma$, $P_v$, and $P_s$). Each iteration simulates a propagation for one video. We calculate the average, maximum and minimum of the cumulated video views ($I_3 + R$) and video shares ($I_3$) along the timeline in Fig. 3.6. As the figure shows, the views and shares are quite diverse for each video—even under the same system settings. This result confirms our earlier measurement results from [4], in which we found that the number of video views and the number of video shares have only very weak correlations with the average share rate ($P_s$) and reception rate ($P_v$).

### 3.2.3  Implications

We use this model to analyze an interesting measurement finding—the limited propagation range—and evaluate our proposed recommendation strategy, which aims to increase the propagation range. The results found by Cheng et al. [1] showed that the sizes of most propagation trees are below 100, and even the most popular videos have relatively small tree sizes as compared with the total number of users in the system. In other words, a vast majority of the cascades vanish quickly. Even the most popular videos do not reach "epidemic" proportions in social networks. This certainly contradicts the expectation that the shared videos will spread as broadly as possible. Moreover, it is counterintuitive because many of the videos imported into social networks are popular in the original video sharing sites.

An underlying reason for the limited spread of videos in social networks is the mechanism for social contagion [5]. According to the current contagion mechanism, only a video shared by a user's friend will appear on the user's page. Videos watched—but not shared—by a user's friends will not appear on the user's news feed. In other words, even if a video is watched by many users, if they do not share the video, the propagation will stop. Unfortunately, according to our statistics, only 16 % of users on average will watch a video shared by a friend; among those, only 13 % will share the video further. Assume that a user has $n$ friends. For a sharer, the expected number of friends who will share this video after watching is thus $n \cdot 0.16 \cdot 0.13$, which we refer to as the epidemic index. When $n$ is less than 48, the epidemic index is less than 1; in this case, the number of sharers diminishes rapidly, and the propagation quickly stops.

Because the number of friends, share rate, and reception rate are intrinsic system properties that cannot be tuned, a practical way to boost propagation is to modify the contagion mechanism; in particular, to leverage the users' viewing information. The simplest solution would be to set up the system so that after a user watches a video, the link for that video would appear in the news feeds of that user's friends. This watching behavior would accurately reflect the popularity of the video among friends and could be even more directly manipulated than the sharing behavior. However, it does not preserve user privacy because the information about every watched video is now distributed to all a user's friends. Therefore, we suggest an anonymous solution: For any user, after a video has been viewed by K friends of that user, the video will appear in the user's news feed as a system-suggested news item, even if none of the friends have shared the video. A possible system-generated comment with the shared video link might be "K friends have viewed this video." There is no need to mention the names of the friends, so that the privacy of other users is well preserved. The key issue for this view-aware contagion strategy is to set the threshold $K$. A small $K$ would be more effective for promoting the propagation, but might trigger excessive news feeds.

## 3.3 Summary

In this chapter, we studied the social propagation process, which determines the social popularity. We presented the observed social propagation patterns, including the social locality, temporal locality, and geographical locality. Based on our observations, we proposed an enhanced epidemic model to capture social propagation, and presented implications based on our propagation model.

# References

1. Xu Cheng, Haitao Li, and Jiangchuan Liu. "Video Sharing Propagation in Social Networks: Measurement, Modeling, and Analysis". In: *IEEE International Conference on Computer Communications (INFOCOM)*. 2013.
2. Daryl J. Daley, Joe Gani, and Joseph Mark Gani. *Epidemic Modelling: An Introduction*. Cambridge Studies in Mathematical Biology. Cambridge University Press, 2001.
3. J. Jiang et al. "Understanding Latent Interactions in Online Social Networks". In: *ACM Internet Measurement Conference (IMC)*. 2010.
4. Haitao Li, Xu Cheng, and Jiangchuan Liu. "Understanding Video Sharing Propagation in Social Networks: Measurement and Analysis". In: *ACM Transactions on Multimedia Computing, Communications, and Applications (TOMM)* 10.4 (2014), p. 33.
5. Greg Ver Steeg, Rumi Ghosh, and Kristina Lerman. "What Stops Social Epidemics?" In: *International Conference on Weblogs and Social Media (ICWSM)*. 2011.
6. Zhi Wang et al. "Propagation-based Social-aware Replication for Social Video Contents". In: *ACM International Conference on Multimedia (Mul- timedia)*. 2012, pp. 29–38.

# Chapter 4
# Propagation-Based Social Video Content Replication

Online social network has reshaped the way multimedia content is generated, distributed, and consumed on today's Internet. Given the massive amount of user-generated content shared through online social networks, users are moving toward accessing this content directly from their preferred social network services. It is intriguing to study the service provision of social content for global users that provide a satisfactory quality of experience (QoE). In this chapter, we present propagation-based social-aware delivery studies.

## 4.1 Inferring Propagation for Social Content Replication

We are facing the following challenges in distributing social content with satis-factory QoE: (1) The huge amount of user-generated content (UGC) requires an equally massive amount of storage and network resources. For example, YouTube has hit a new record, with 100 h worth of videos uploaded by users every minute; (2) Newly generated content tends to attract most of the users, but it is difficult to estimate its popularity for the purposes of allocating services properly—a task that is dynamically affected by social networks themselves [3]; (3) Social content has close-to-uniform [6] but highly volatile popularity profiles because a large portion of that content is shared among small social groups (e.g., family members).

Challenge (1) makes traditional service paradigms (e.g., C/S based on private servers) unsuitable because it is too expensive to replicate all content to all servers. Instead, a common practice for providing these content services is to replicate content to servers in different geographic regions [1] by allocating resources from the geo-distributed Content Delivery Network (CDN) or cloud, where content can be dynamically distributed to serve users all over the world. Challenges (2) and (3) make the traditional replication approaches, which work well only for content with a skewed and stable popularity profile, unsuitable in the context of online social

© The Author(s) 2016
Z. Wang et al., *Social Video Content Delivery*, SpringerBriefs in Electrical
and Computer Engineering, DOI 10.1007/978-3-319-33652-7_4

networks. Mislove et al. [7] observed a large reduction in the cache hit ratio when traditional caching schemes were used to replicate social content.

Our previous study [9] revealed the key observation that social content, unlike regular content, does not propagate among users randomly—instead, such content propagates along the social-network topology based on several rules of social propagation, due to the social behaviors including posting and re-sharing content.

Based on the social popularity and propagation investigated in the previous chapters, we will present a general framework with propagation patterns and predictions incorporated. Then, we develop a social-aware delivery system to effectively distribute social content with superb QoE. In this chapter, we use the most representative type of multimedia content—social videos—to investigate how social content can be effectively replicated based on social propagation. However, our design can be used to deliver a variety of multimedia types.

Based on the propagation patterns, we will also study propagation predictors to guide content delivery. In particular, the propagation region predictor, global audience predictor, and local audience predictor answer the following questions, respectively: (a) Which videos should be replicated to which edge-cloud servers? (b) How much bandwidth should be reserved for each video by the edge-cloud? and (c) Which videos should be served by which peers?

Furthermore, we will present a propagation-based social content delivery framework that employs a hybrid edge-cloud and peer-assisted video replication architecture. Based on the propagation predictions, videos are replicated by both the edge-cloud servers and peers at different geographic locations as follows: (1) We design the edge-cloud replication strategies according to the region predictor and global audience predictor to determine the region selection and bandwidth reservation; (2) We further design peer-assisted replication to function according to the local audience predictor, performing social-aware cache replacement at each peer.

## 4.2  Edge-Network Replication Architecture

According to our observations, users involved in social video propagation are socially and geographically close to each other, and their social actions are clustered over a short period of time. Accordingly, we propose a *hybrid edge-cloud and peer-assisted* architecture for social video delivery. In this architecture, the edge-cloud can support the time-varying bandwidth and storage allocations requested by different regions, while the peers are able to contribute to each other in similar social groups. Figure 4.1 illustrates the conceptual architecture of our design, in which two overlays are presented as follows: (1) social propagation overlay based on the social graph, which determines the video propagation among friends (i.e., which users can share a generated video with their direct friends) who may further re-share the video to more people, and (2) delivery overlay, which determines how video content is delivered from edge-cloud servers to users or among users in a P2P paradigm.

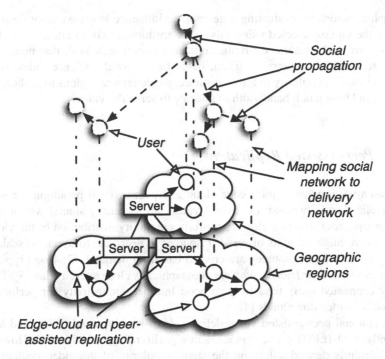

**Fig. 4.1** Conceptual architecture of the propagation-based social-aware content replication system

This architecture not only makes use of the edge-cloud servers distributed among different geographic regions to serve social videos to users in different regions, but also schedules peers to cache the video content in their local storage so that they can help each other download the videos.

In designing our Propagation-based Social-Aware Replication (PSAR) for social content delivery, we will study edge-cloud replication, which controls how videos are replicated to edge-cloud servers as well as peer-assisted replication, which controls how videos are cached at peers.

## 4.2.1   Edge-Cloud Replication

In edge-cloud video replication, videos are generally replicated to servers located in different geographic regions. The main purpose of edge-cloud replication is to allow users in different regions to download the videos they request from their local servers (which are located in the same regions as the users) to improve the quality of video service [2].

We redesign the edge-cloud replication strategy by taking social propagation into account. We first select the videos that are the most likely to propagate across geographic regions by evaluating a geographic influence index we have designed. Because the videos selected using this index are more likely to attract users from more regions in the future, we replicate them to more regions so that more users will be served by local servers. Then, based on the local audience index, which reflects a video's popularity in the near future, we determine regions to replicate the video to, and how much bandwidth to allocate to serve the video.

### 4.2.2   Peer-Assisted Replication

The reason we propose a joint edge-cloud and peer-assisted paradigm for social video replication is twofold: (1) Social videos are generally shared within small social groups, resulting in a close-to-uniform popularity distribution of the videos, which require huge amounts of server resources to distribute to users. To scale the delivery system, peer resources are created based on demand. (2) Users typically share videos with their friends, who are geographically close to each other [8]. These socially connected users tend to have good Internet connectivity for performing peer-assisted video downloads [4].

In traditional peer-assisted video delivery, least-recently used (LRU) and least-frequently used (LFU) cache replacement algorithms are widely used. However, such algorithms depend solely on the static popularity of the video content and cannot achieve good performance when the access patterns of videos are affected by social activities in the online social network. Based on the local audience index summarized from the propagation pattern, we also redesign the peer cache replacement algorithm. Specifically, we let peers cache videos that not only improve the peer contribution, i.e., caching these videos can improve the fraction of the video content uploaded by peers over that uploaded by both peers and servers, but also improve the possibility of peers serving unpopular videos to their local friends. In turn, these friends benefit from their good Internet connectivity to the local peers.

### 4.2.3   Design Challenges

In PSAR, the replication of social video content faces great resource-allocation challenges in the presence of multiple video propagations. Figure 4.2 illustrates an example in which there are only two videos. In this figure, the circles represent users who are located in different geographic regions (e.g., region 1 and region 2). User $A$ generates and shares video $a$ in timeslot $T$. That video is re-shared by $A$'s friends $C$ and $D$ in timeslot $T+1$. Concurrently, another user, $B$, generates a different video, $b$. Both video $a$ and video $b$ will propagate across the social connections, and the two propagation trees may intersect in the same region or at the same peer. In other words, both region 1 and region 2 are involved in the two propagation trees, and both videos may reach user $K$ in timeslot $T+3$. The resource allocation must determine

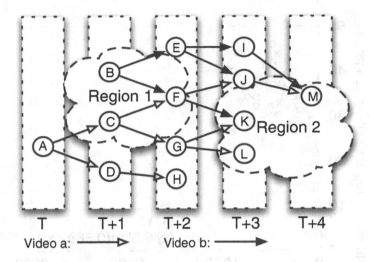

**Fig. 4.2** Resource allocation for two propagation trees

(1) how to serve video $a$ and $b$ using the edge-cloud servers in regions 1 and 2, and (2) how to cache the videos $a$ and $b$ at peer locations to improve downloads for others. This is a huge challenge when many videos are propagating at the same time. Therefore, we will discuss the two problems separately for edge-cloud replication and for peer-assisted replication, respectively.

## 4.3 Propagation Prediction for Replication

In this section, we establish the connection between social video propagation and video replication, using the propagation prediction.

### 4.3.1 Propagation Region Prediction

Based on the dataset used in our measurement studies, Fig. 4.3 illustrates the correlation between the number of regions involved in the video propagation and the propagation size for different videos. Notably, a large propagation size generally results in more regions (city-level locations) being involved in the propagation. In particular, the relationship follows a logarithmic function. In PSAR, the propagation size is utilized to determine whether a video should be replicated to more regions. In particular, we design a geographic influence index as follows:

$$g_v^{(T)} = c_1 \log(c_2 s_v^{(T-1)}), \tag{4.1}$$

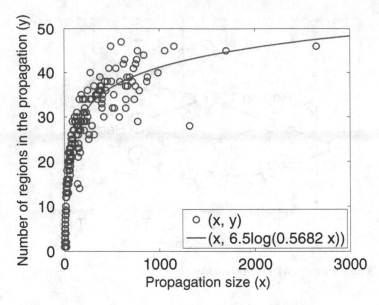

**Fig. 4.3** Number of regions in the propagation versus the size of the propagation

where $s_v^{(T-1)}$ is the propagation size of the propagation tree of video $v$ in timeslot $T-1$. A large $g_v^{(T)}$ value indicates that more regions will be involved in the propagation of the video. Intuitively, a video should be replicated to more regions when the predicted number of regions involved in the propagation is larger than the number of regions to which it has already been replicated.

### 4.3.2 Global Audience Prediction

To allocate bandwidth to serve a social video content, we design a global audience predictor, based on a global audience index to evaluate the strength of a video's propagation in timeslot $T$, using the propagation information as follows: (1) the current propagation size ($s_v^{(T)}$); (2) the current propagation depth ($h_v^{(T)}$); and (3) the time lag since the propagation tree was formed ($\tau_v^{(T)}$). The global audience index is defined as follows:

$$e_v^{(T)} = z_s(\tau_v^{(T)})(s_v^{(T)}/h_v^{(T)}), \qquad (4.2)$$

where $z_s(\tau_v^{(T)})$ is a decreasing function to make use of the temporal locality that can adjust the global audience index according to $\tau_v^{(T)}$: more recently, generated or shared videos will have a larger global audience index. Based on our observation in Chap. 3, $z_s(t)$ is defined as follows:

$$z_s(t) = 1/(t^s \sum_{k=1}^{N} \frac{1}{k^s}), \qquad (4.3)$$

where $s$ is the zipf shape parameter and $N$ is the number of hours between the publication time of the earliest video and the publication time of the latest video. In our design, $e_v^{(T)}$ will be used to guide the replication. Larger $e_v^{(T)}$ values indicate that more users can join the propagation tree in timeslot $T$. The rationale for $e_v^{(T)}$ is as follows: (1) A larger $s_v^{(T)}$ indicates that more users can be reached by the video, and these users are the potential viewers (downloaders) of video $v$; (2) According to the social locality, a small $h_v^{(T)}$ indicates that users in the propagation tree are still social-aware to the root user; therefore, the video can still reach more users; (3) According to the temporal locality, a large $\tau_v^{(T)}$ slows down the propagation. Based on the global audience index, we can determine how much bandwidth we need to reserve for a video in a future timeslot in PSAR.

Figure 4.4 compares our social-aware global audience prediction and the traditional popularity estimation using only the historical popularity. The effectiveness of our global audience prediction is verified as follows. In Fig. 4.4a, each sample represents a video's current popularity versus its popularity in the previous timeslot. We observe that the video's global audience is highly volatile over time, with only a very small correlation between the current size of the audience and the previous size of the audience. Our prediction is illustrated in Fig. 4.4b. Each sample represents the current popularity versus the global audience index for the previous timeslot. After incorporating the propagation patterns, the correlation coefficient is four times larger, indicating that future popularity can be better predicted by our design.

### 4.3.3 Local Audience Prediction

In our architecture, a peer performs the cache replacement locally using not only the perceived video's popularity but also local social factors. To determine which videos should be stored at a peer, we design a local audience predictor based on the following information at peer $i$: (1) the local popularity, which is the number of requests peer $i$ receives for video $v$, denoted as $p_i^v$; (2) the fraction of peer $i$'s friends that can join the propagation tree of video $v$, denoted as $f_i^v$. The $f_i^v$ value is calculated by historical records for different video categories, i.e., peer $i$ keeps a record of the fraction of friends that have been attracted by each category in its history; and (3) the time lag between the time when a propagation tree is constructed and the time when the peer re-shares the video, denoted by $\tau_v^{(T)}$. Based on the social propagation patterns, we design a local audience index to perform the prediction as follows:

$$q_v = z_s(\tau_v^{(T)})(p_i^v f_i^v). \qquad (4.4)$$

**Fig. 4.4** Global audience prediction. (**a**) New popularity versus historical popularity. (**b**) New popularity versus the global audience index

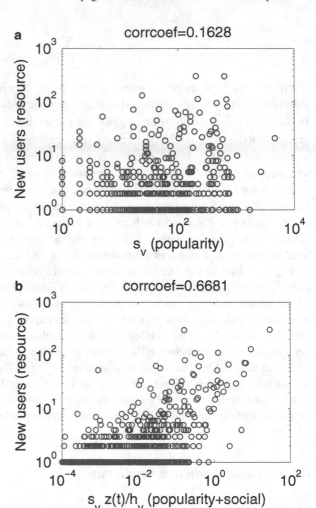

In peer-assisted replication, videos with a smaller local audience index are more likely to be dumped by the peer. The rationale is that a larger $p_i^v f_i^v$ indicates that peer $i$ can potentially attract more users to re-share video $v$ from its friends in the future, and $\tau_v^{(T)}$ is utilized to reflect the temporal locality.

The effectiveness of the local audience predictor is verified by our data as well. Figure 4.5 illustrates the CDF of the correlation coefficient between a friend's video category preference (calculated as a category preference vector) at time $T$ and the category preference at time $T - 1$. We observe that most of the friends' preferences can be inferred from their historical preference. In our dataset, the correlation coefficient for 80 % of the user preference measures [9] in two consecutive timeslots can be larger than 0.8.

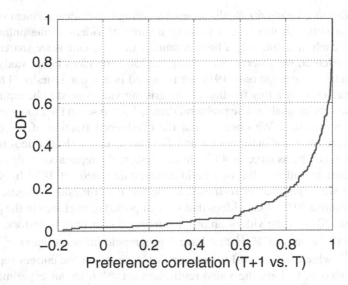

**Fig. 4.5** CDF of preference of local friends

## 4.4 Propagation-Based Replication Strategies

### 4.4.1 Region Selection in Edge-Cloud Replication

Based on the region prediction and the global audience prediction, we first select the videos to be replicated and determine which regions they should be replicated to, and, then, we reserve upload bandwidth at edge-cloud servers for these videos. When performing video replication, we must also discover the videos that may propagate to additional regions in the future. We use the geographic influence index to perform region prediction for that purpose. To achieve better video download quality, a video with a larger $g_v^{(T)}$ should be replicated to more regions to serve users locally. Parameters $c_1$ and $c_2$ are selected based on this measurement. Based on the geographic influence index, we can predict whether the regions to which the video has currently been replicated are sufficient.

*Initial Replication* After video $v$ is first generated by a user in the online social network, it will be stored by the server that is closest to that user's friends. Let $d_{r,i}, i \in \mathscr{F}_v$ denote the geographic distance between region $r$ and user $i$, where $\mathscr{F}_v$ is the set of friends of the root user of video $v$ ("distance" based on an Internet connectivity measurement can also be used, e.g., bandwidth or RTT). The initial region is then selected by solving the following equation: $r_v = \arg\min_{r \in \mathscr{R}} \sum_{i \in \mathscr{F}_v} d_{r,i}$, where $\mathscr{R}$ is the set of regions that can be used for the replication (determined by the cloud providers), and $r_v$ is the region selected for the replication.

*Selecting Existing Videos for Replication* According to our measurement study, we observe that although there are a massive number of videos in the online social network, in each timeslot, only a limited number of those videos are shared among users. In particular, we observe that among the 350,860 videos in our study, by our measurements, on average only 1919 are re-shared in any one timeslot (1 h). Thus, in each timeslot, only a tiny fraction of the existing videos need to be replicated to improve the service quality. The problem, then, is how to select the proper candidate videos for replication. We observe that the overlapped fraction of the common videos that are re-shared in timeslot $T$ and $T-1$ compared with all videos re-shared in timeslot $T$ can be as large as 49 %. In our design, the replication video set $\mathcal{V}^{(T)}$ is constructed as follows. We first build a candidate video set $\mathcal{W}^{(T)}$ by selecting videos that were imported or re-shared in the previous timeslot. In particular, we randomly choose 80 % of the videos that were imported or re-shared in the previous timeslot and 20 % of the videos among the most popular ones in history. Second, we choose the videos in $\mathcal{W}^{(T)}$ that have a geographic influence index $g_v^{(T)}$ larger than $\theta_v^{(T+1)}$, which is a control parameter that depends on the current replication status of video $v$, to form the video replication set $\mathcal{V}^{(T)}$. In our experiments, we let $\theta_v^{(T)} = 0.8|\mathcal{R}_v^{(T)}|$, where $\mathcal{R}_v^{(T)}$ is the set of regions to which $v$ has already been replicated. The rationale is that a video should be replicated to more regions if its current replication status is below the requirement estimated from the geographic influence index.

*Selecting Replication Regions for Videos in* $\mathcal{V}^{(T)}$ After constructing $\mathcal{V}^{(T)}$, the videos in $\mathcal{V}^{(T)}$ must be replicated to more regions. Because these videos are the candidates that can attract users from different regions, we must determine which videos need to be replicated to which regions. In our design, we extend the replication of a video to one additional region each time. The selection of the new region is similar to the approach for selecting the initial region. We minimize the geographic distance between the region and the potential users who may join the propagation tree. Let $\mathcal{L}_v^{(T)}$ denote the set of users who joined the propagation tree in the previous timeslot. Then, the selection is made as follows:

$$r_v = \arg \min_{r \in \mathcal{R} - \mathcal{R}_v^{(T)}} \sum_{i \in \bigcup_{k \in \mathcal{L}_v^{(T)}} \mathcal{F}_k} d_{r,i}, \qquad (4.5)$$

where $\mathcal{F}_k$ is the friend set of user $k$. The rationale behind this approach is that because the users in $\mathcal{L}_v^{(T)}$ are those who joined the propagation tree in the previous timeslot, it is likely that they will attract new video users due to the temporal locality of propagation. We utilize these users' friends' locations as a sample of all the users that can join the propagation tree. Then, we select the region that is closest to all those users. The benefit of always extending a video to a new region in the replication (i.e., $r_v$ is selected from $\mathcal{R} - \mathcal{R}_v^{(T)}$) is that users in a popular propagation tree are able to choose from among more regions from which to download the video content. Moreover, our scheme improves the possibility that they will select the preferred regions.

### 4.4.2 Bandwidth Reservation for Social Content at Edge-Cloud Servers

During each scheduled iteration, we allocate upload bandwidth at the edge-cloud servers for the replicated videos. In our design, the amount of bandwidth reserved depends on the social propagation strength, which can be evaluated by the global audience index $e_v^{(T)}$. Let $\mathcal{V}_r$ denote the set of videos that are replicated in region $r$. The bandwidth reservation is then performed as follows:

$$b_{v,r_v} = B_{r_v} e_v^{(T)} / \sum_{v \in \mathcal{V}_{r_v}} e_v^{(T)}, \forall v \in \mathcal{V}^{(T)}, \tag{4.6}$$

where $b_{v,r_v}$ is the amount of bandwidth to be reserved for video $v$ in the selected replication region $r_v$ when that region is fully loaded with requests. A video can be extended to use more bandwidth than $b_{v,r_v}$ when the region is not fully loaded. $B_r$ is the upload capacity of region $r$. The rationale for the bandwidth reservation is that videos with larger $e_v^{(T)}$ tend to attract more users in the propagation in the near future and, therefore, more upload bandwidth should be allocated for these videos' propagation to benefit the potential downloaders. Our edge-cloud replication algorithm is illustrated in Algorithm 1.

---

**Algorithm 1** Edge-Cloud Replication Algorithm.

---

1: **procedure** VIDEO AND REGION SELECTION
2:      $\mathcal{V}^{(T)} \leftarrow \Phi$
3:      **if** $v$ is newly published **then**
4:          $\mathcal{V}^{(T)} \leftarrow \mathcal{V}^{(T)} \cup \{v\}$
5:          $r_v \leftarrow \arg\min_{r \in \mathcal{R}} \sum_{i \in \mathcal{F}_v} d_{r,i}$
6:      **else**
7:          **if** $v \in \mathcal{W}^{(T)}$ and $g_v^{(T)} > \theta_v^{(T+1)}$ **then**
8:              $\mathcal{V}^{(T)} \leftarrow \mathcal{V}^{(T)} \cup \{v\}$
9:              $r_v \leftarrow \arg\min_{r \in \mathcal{R} - \mathcal{R}_v^{(T)}} \sum_{i \in \bigcup_{k \in \mathcal{L}_v^{(T)}} \mathcal{F}_k} d_{r,i}$
10:         **end if**
11:      **end if**
12: **end procedure**
13: **procedure** BANDWIDTH RESERVATION
14:      **for all** $v \in \mathcal{V}^{(T)}$ **do**
15:          **if** $v$ is replicated at region $r_v$ **then**
16:              $b_{v,r_v} \leftarrow B_{r_v} e_v^{(T)} / \sum_{v \in \mathcal{V}_{r_v}} e_v^{(T)}$
17:         **end if**
18:      **end for**
19: **end procedure**

---

### 4.4.3 Cache Replacement in Peer-Assisted Replication

We have shown why the unique propagation patterns in social networks make it extremely attractive to utilize a peer-assisted paradigm, which allocates a certain amount of resources from the users to replicate the video content: Peers (users) are likely to be able to serve their social neighbors with good Internet connectivity. In our peer-assisted replication, we assume that users download video content according to their individual demands. We then design a social-aware cache replacement strategy for peers to determine which videos are cached to help other users. The caching strategy used for peers can greatly affect the performance of a P2P system [5].

*Peer Cache Replacement* A large local audience index indicates that the video can potentially be downloaded by a great number of local friends; therefore, the peer should keep the video to serve those friends. Thus, in our cache replacement algorithm, the peer will try to dump cached videos with the smallest $q_v$'s until it has sufficient capacity to cache new videos.

## 4.5 Summary

We designed a propagation-based social-aware content delivery framework using a data-driven approach. Based on known propagation patterns, which demonstrate social, geographical, and temporal localities, we created propagation predictors that enable propagation-based social-aware replication strategies to serve social content to users. Specifically, we proposed three replication indices: a geographic influence index, a global audience index, and a local audience index that guide region selection, bandwidth reservation, and cache replacement, respectively, in the proposed joint edge-cloud and peer-assisted replication framework.

## References

1. V.K. Adhikari, S. Jain, and Z.L. Zhang. "Where Do You Tube? Uncovering YouTube Server Selection Strategy". In: *IEEE International Conference on Computer Communications and Networks*. 2011.
2. V.K. Adhikari et al. "Reverse Engineering the YouTube Video Delivery Cloud". In: *IEEE Hot Topics in Media Delivery Workshop*. 2011.
3. M. Cha, A. Mislove, and K.P. Gummadi. "A Measurement-Driven Analysis of Information Propagation in the Flickr Social Network". In: *ACM International Conference on World Wide Web (WWW)*. 2009, pp. 721–730.
4. B. Huffaker et al. "Distance Metrics in the Internet". In: *IEEE International Telecommunications Symposium*. 2002.
5. J.G. Luo et al. "A Trace-Driven Approach to Evaluate the Scalability of P2P-Based Video-on-Demand Service". In: *IEEE Transactions on Parallel and Distributed Systems (TPDS)* 20.1 (2009), pp. 59–70.

6. A. Mislove. "Rethinking Web Content Distribution in the Social Media Era". In: *NSF Workshop on Social Networks and Mobility in the Cloud*. 2012.
7. A. Mislove et al. "You Are Who You Know: Inferring User Profiles in Online Social Networks". In: *ACM International Conference on Web Search and Data Mining (WSDM)*. 2010.
8. S. Scellato et al. "Distance Matters: Geo-Social Metrics for Online Social Networks". In: *USENIX Conference on Online social networks*. 2010.
9. Zhi Wang et al. "Propagation-based Social-aware Replication for Social Video Contents". In: *ACM International Conference on Multimedia (Mul- timedia)*. 2012, pp. 29–38.

# Chapter 5
# Concluding Remarks

With the advances in online social networking, users rather than content providers determine how videos reach other users. In this book, we surveyed recent research on social video delivery with the goal of improving the quality of experience for massive numbers of users. We identified the unique patterns and characteristics of social video propagation and content popularity through large-scale log analyses. We demonstrated a series of strategies with great potential, including content replication, crowdsourced content caching, and network resource allocation. This new and promising research area has many challenging issues that need to be addressed in the near future—and we believe that a data-driven and social-aware framework design will be a key part of the solution. Within this framework, a deep understanding of user behavior, social propagation structures, knowledge of content characteristics and context information, as well as social-relationship-based collaborative content sharing mechanisms must be developed because all these factors will play important roles.

Given the unique characteristics of social videos, a series of solutions have been proposed in the literature, addressing various aspects of the challenges. We believe the following research directions are worth further investigation.

**Geo-Distributed Cloud for Social Video Distribution** Geographical information becomes useful as we move toward cloud platforms. Many new generations of cloud-based multimedia services have emerged in recent years, and these services are rapidly changing both operational and business models in the marketplace. A prominent example is Netflix[1] a major on-demand Internet video provider, which migrated its entire infrastructure to the powerful Amazon Web Services (AWS) cloud in 2012, using Elastic Cloud Computing (EC2) to transcode master video copies and the Simple Storage Service (S3) for content storage. Several studies have addressed using cloud resources for social content delivery. Wu et al. [2] considered

---

[1]Netflix: http://www.netflix.com.

© The Author(s) 2016
Z. Wang et al., *Social Video Content Delivery*, SpringerBriefs in Electrical and Computer Engineering, DOI 10.1007/978-3-319-33652-7_5

a generic geo-distributed cloud infrastructure that consists of multiple *cloud sites* distributed across different geographical locations. Using the geo-distributed cloud resources, social video content can be replicated to locations closer to users.

**Instant Social Video Delivery**  With the rapid development of mobile networking and end-terminals, anytime and anywhere data access has become readily available. Given the capability for crowdsourced content capture and sharing, the duration users expect videos to have has become shorter and shorter. Representative sites such as Twitter's Vine that are available exclusively to mobile users enable users to create ultra-short video clips and to instantly post and share them with their followers. Taking Vine as a case study [4], Zhang et al. presented evidence that instant social videos have a short lifetime and a highly skewed popularity that decays quickly over time. Videos in these trending social media sites are both more fragmented and instantaneous—driven by the paradigm shift to mobile and cloud computing. The results indicate that a middleware framework integrated with a pre-fetching and viewing-time scheduling scheme is promising for providing improved quality of experience.

**Mobile Social Video Delivery**  Mobile caching, by replicating bandwidth-intensive videos on edge-network devices (e.g., users' smartphones) is becoming promising for social video delivery. Previous studies have demonstrated that such device-to-device (D2D) content sharing is possible when users are close to each other, and the content to be delivered by users can tolerate delays [1]. However, in traditional D2D content sharing, a user broadcasts generated content or re-shared content to a set of random users that are close by. As a result, all content is disseminated to users in the same way (e.g., random flooding), causing the following problems: (1) In greedy flooding, smart devices in edge networks have to expend precious power to cache and relay large amounts of content. As the number of user-generated social content increases, such broadcasting mechanisms are inherently non-scalable; (2) Social videos have heterogeneous popularity, while the conventional approaches treat them all the same, leading to wasted resources carrying unpopular content; (3) Because of dynamic mobility patterns, users may not be able to fetch content timely.

To address these issues, a joint propagation- and mobility-aware device-to-device replication strategy can be developed based on social propagation characteristics and crowd mobility patterns in edge-network *regions*, e.g., an area of hundred-meter range that users can move across. As illustrated in Fig. 5.1, using the social graph and propagation patterns, we first estimate how content will be received by users, and we then predict which regions users will be moving within and how long they will stay. Rather than flooding content between users that are merely close to each other, we disseminate social videos according to the influence of users and the propagation of videos. In Fig. 5.1, for example, user $e$—while not a friend of any other user—is moving to the region where users $c$ and $d$ are. Thus, $e$ will be selected to replicate the content generated by user $a$, and users $c$ and $d$ will receive the content shared by user $a$ at times $T2$ and $T3$, respectively.

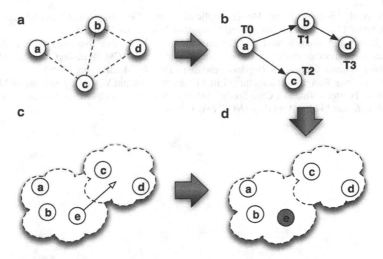

**Fig. 5.1** Device-to-device replication affected by social topology, content propagation, and user regional mobility. (**a**) Social topology. (**b**) Content propagation. (**c**) User regional mobility. (**d**) Off-grid replication assignment

**Crowdsourced Interactive Live Social Streaming** Empowered by today's rich tools for media generation and collaborative production, the multimedia service paradigm has been shifting from a conventional single source, to multi-source, to many sources, and now, toward crowdsourcing. Crowdsourced live streaming platforms such as Twitch.tv allow general users to broadcast their content to massive numbers of viewers, thereby greatly expanding the content and user bases. However, the resources available for these non-professional broadcasters are limited and unstable, which can potentially impair the streaming quality and affect viewers' experiences. The diversity of live interactions among the broadcasters and viewers can further aggravate the problem. Zhang et al. [3] presented an initial investigation on modern crowdsourced live streaming systems. Taking Twitch as a representative, they revealed that the view patterns are determined both by events and by broadcasters' sources. The current delay strategy on the viewer's side substantially impacts the viewers' interactive experience, and there is a significant disparity between the long broadcast latency and the short live messaging latency. On the broadcaster's side, dynamic uploading capacity is a critical challenge that noticeably affects the smoothness of live streaming for viewers.

# References

1. Xiaofei Wang et al. "TOSS: Traffic offloading by social network service-based opportunistic sharing in mobile social networks". In: *IEEE International Conference on Computer Communications (INFOCOM)*. 2014.

2. Yu Wu et al. "Scaling Social Media Applications into Geo-Distributed Clouds". In: *IEEE International Conference on Distributed Computing Systems (ICDCS)*. 2012.
3. Cong Zhang and Jiangchuan Liu. "On crowdsourced interactive live streaming: a Twitch. tv-based measurement study". In: *Proceedings of the 25th ACM Workshop on Network and Operating Systems Support for Digital Audio and Video*. ACM. 2015, pp. 55–60.
4. Lei Zhang, Feng Wang, and Jiangchuan Liu. "Understand Instant Video Clip Sharing on Mobile Platforms: Twitter's Vine as a Case Study". In: *ACM Network and Operating System Support on Digital Audio and Video Workshop (NOSSDAV)*. 2014.

Printed in the United States
By Bookmasters

Printed in the United States
By Bookmasters